2 MOUNT MCKINLEY FROM WONDER LAKE

4   BULL MOOSE, KENAI PENINSULA

BLACK-LEGGED KITTIWAKES AND COMMON MURRES, CAPE NEWENHAM

JOHNS HOPKINS BAY AND GLACIER, GLACIER BAY

Photography Credits

All photographs by Steve Kaufman or Yogi Kaufman in approximately equal numbers, except for the photograph on page 130-131 which is by Mary Bee Kaufman (Mrs. Steve Kaufman).

Excerpt from "The Spell of the Yukon" in *The Collected Poems of Robert Service.* Reprinted by permission of Dodd, Mead & Company, Inc. Excerpt from *Travels in Alaska* by John Muir. Reprinted by permission of Houghton Mifflin Company.

Published by Lickle Publishing Inc
590 Madison Avenue, New York, NY 10022

Copyright © 1987, 1997 by Lickle Publishing Inc
Photographs copyright © 1987, 1997 by Steve Kaufman,
Yogi Kaufman, and Mary Bee Kaufman
Introduction copyright © 1987 by Margaret E. Murie
Diversity and Epilogue copyright © 1997 by Yogi, Inc

All rights reserved. No part of this publication may be reproduced, stored in a retrieval system, or transmitted in any form or by any means—electronic, mechanical, photocopying, recording, or otherwise-without prior written permission from the publisher.

Library of Congress Cataloging-in-Publication Data

Kaufman, Steve.
    Untamed Alaska / photography by Steve and Yogi Kaufman;
introduction by Margaret E. Murie.
        p. cm.
    ISBN 1-890674-00-1
    1. Natural history — Alaska — Pictorial works. 2. Zoology — Alaska —
Pictorial works. 3. Alaska — Pictorial works. I. Kaufman, Yogi. II. Title.
QH105.A4K433   1997
508.798'022'2-dc21   97-16755
                                                                CIP

Opposite:
BRAIDED GLACIAL MELT PATTERNS, POLYCHROME PASS

Printed in Japan by Toppan Printing Co., Inc.

10TH ANNIVERSARY EDITION

UNTAMED ALASKA

Photographs by Steve and Yogi Kaufman
Introduction by Margaret E. Murie

LICKLE PUBLISHING INC

BROWN BEAR, ALASKA PENINSULA

Whenever I give a lecture on my home country Alaska, I always find the first difficulty is communicating a sense of its size. I can say to my audience that Alaska's area is 586,000 square miles, larger than Norway, Sweden, and Finland combined; that its shoreline is 34,000 miles long; that its lakes cover an estimated 60 million acres; that one of its glaciers is larger than Rhode Island. I can say that it has six major rivers which I am glad have all kept their native names: Yukon, Noatak, Kobuk, Kuskokwim, Tanana, and Koyukuk.

I can add that Alaska is crossed, west to east, by three great mountain chains, the coastal mountains and the Alaska and Brooks Ranges. This accounts for the contrast in climate—the very wet coast, the dry Interior. Fairbanks, in the middle of the state, has one of the greatest temperature ranges of any place in the world, from 60° below in the winter to 98° above in summer!

Just north of Anchorage are the farms of the famous Matanuska Valley, with their huge strawberries and 30-pound cabbages. Some of the largest fishing operations in the world flourish in the Bristol Bay region and on through the Aleutians to Kodiak Island, site of the second-ranking fishing port under the American flag. Fishing continues on southward through Prince William Sound and down the southeast coast to the great forests. There are also broad interior valleys, at least three still-active volcanoes, and hundreds of islands—the Pribilofs, the Aleutians, and those of the Alexander Archipelago.

And when I have said all that, I still don't know what kind of picture lodges in the minds of my audience. In 1908, Ella Higginson said, "No writer has ever described Alaska; no writer ever will." I realize now what she meant.

The photographs in this book will give you an idea of the natural beauty and amazing wildlife in our 49th state. The joyous fact is that nearly all of the species present when the white man came are still here, still in their age-old habitats. The natives who were here in 1741 when Bering discovered that Alaska was entirely separate from Asia lived in the wilderness without destroying it. For this we can be eternally grateful.

We can also be grateful for the wisdom and foresight of William H. Seward, secretary of state, and the courage of Andrew Johnson, that much-maligned president. Thanks to them, Alaska was purchased from Russia in 1867. (The czar needed money for European wars. Apparently the Russians felt that they had already obtained most of Alaska's treasures in the pelts of sea otters and fur seals.) But we cannot be proud of the next few years in the state's history. Congress and most of the people seemed to have forgotten Alaska. Those who remembered it at all engaged in lamenting its purchase and decrying its worth. The Navy, the Army, the Bureau of Education, and the collector of customs comprised in a restricted and haphazard manner what might be called a temporary government. It took years

"The joyous fact is that nearly all of the species present when the white man came are still here, still in their age-old habitats."

of agitation before Congress finally passed the Organic Act providing a civil government for Alaska, signed into law by President Chester A. Arthur on May 17, 1884. Statehood was not granted until 1959. And in those intervening years were tragedies and frustrations and dreadful injustices toward the native people. There was also the influx of white men looking for gold and fur, pouring into the Canadian Klondike in 1897. By 1900, they had reached the sands of Nome, by 1902, the creeks around Fairbanks.

When my family moved to Fairbanks in 1911, the federal government was well established. The Territory had been divided for governmental purposes into four judicial divisions, each with a district judge, U.S. marshal, and land office. Fairbanks, with a population of about 5,000, was the center of the Fourth Division; my stepfather was an assistant district attorney. Reaching to the Arctic coast, the Fourth Division covered 220,000 square miles!

Here in the middle of this enormous land was a little, busy, booming community where mining and trapping were the bases of life. If a miner had a prospect but no funds for food and tools to continue searching, a merchant grubstaked him. If the miner "struck it," the merchant was repaid and typically had a future interest in the mine. And there were usually some lawyers mixed up in all this, too.

Fairbanks had both a grabbing, lusty, frontier manner and a striving for some of the "Outside" way of life. It was torn between the "improper" and "proper" cultures. Front Street on the river had 23 saloons along its four- or five-block length, and in the very middle of town, only two blocks from the federal courthouse, was the red-light district—something to make nine-year-olds question, "Why does that part of town have a board fence around it?"

The respectable women set the pattern for the town and its homes. Steady contact with other people through all kinds of social life—school, library, church—a regular routine, and a definite project for each day helped. It was their bulwark against the isolation, the cold, and the difficulties of housekeeping.

Wash day was Monday. First the yellow card went up in the kitchen window to tell "Fred the Waterman" to bring in extra buckets from his great wooden tank-sleigh drawn by two huge grey horses. We children were warned to stay in the living room by the stove so we would not catch a chill while he carried in all that water and poured it into the big barrel in the corner of the kitchen. Then Mother stoked the kitchen range to heat water in the copper wash boiler, set up washtubs in front of the stove, put the washboard in place, cut the Fels-Naphtha soap into shavings, and melted it for the wash water. And when all the scrubbing and rinsing were done, the clothes had to be hung on lines fastened under the ceiling, for hanging laundry outdoors was impossible; it would freeze immediately. The other days of the week were happiness compared to Monday.

Alaska then was the dancingest place in the world, I think. There were dances every week and many special balls with midnight suppers put on by the lodges. In satin, lace, crêpe de Chine, and correct dark suits, the women and men of Fairbanks would dance all night (Billy Root's orchestra never seemed to tire), and go to the Model Cafe for breakfast. Then they would go home, change clothes, and go to church or to work. Yes, the Masons, Moose, Pioneers of Alaska, Arctic Brotherhood, Eagles, Odd Fellows, and Elk certainly kept social life sparkling.

As for us children, we were few in comparison to adults, since that frontier population was largely made up of unattached men, and so we were pampered. On Saturdays we were all over town, going on errands and racing our Huskies hitched to little coaster sleds. We took for granted the jovial greetings, the help in untangling harnesses, the 50-cent pieces (a quarter was the smallest coin in our town) thrust into our palms: "Here, go buy yourself some candy."

We lived in an atmosphere of tolerance and love. I think nearly every household in town had some miner or trapper friend who became a sort of member of the family. They came in from the creeks for part of the winter and were always there for Thanksgiving and Christmas — and what Santa Clauses they were!

Summer was busy with berry picking, gardening, and trips on the river or out to the creeks to visit miner friends and watch them work. And always there was the big parade on the longest day of the year and the big event, a midnight ball game in that never-failing, warm sunlight. And yes, I must be honest, I suppose — there were mosquitoes, too!

Like every Alaskan town in those days, Fairbanks had citizens of a variety that I now look back on with fascination as well as affection. The love of adventure, the craving for gold, hard times or dull lives in their home place, whatever the circumstances which brought people to the country, they were such that we had every kind from dukes to roustabouts, a wealth of nationalities, professions, and skills. All were treated alike and with a great deal of humor.

Those who distinguished themselves for better or worse earned some interesting nicknames. I knew about the "Blue Parka Man" who was a thief, the "Blueberry Kid" who was so fond of that berry, and the "Seventy-Mile Kid" who later was to guide Archdeacon Stuck on the earliest ascent of Mount McKinley's south peak and afterward become the first superintendent of Denali National Park. I knew "Eat-Em-Up Frank" who ran a roadhouse and always thus announced meal time. The fiery editor of our local newspaper was lame and known as "Step-and-a-Half Thompson."

I think nearly every little town had its orchestra, dance band, and theatre group. All of the churches had choirs. The people of Alaska made their own exuberant way of life. Cold and dark could not defeat them. It was not easy to get into that country — it was not easy to get out of it. Surely, when people reached whatever promising creek they chose and built there some kind of settlement, they could

"The people of Alaska made their own exuberant way of life. Cold and dark could not defeat them. It was not easy to get into that country — it was not easy to get out of it."

well begin to create a livable and lovable community.

I grew up in the Fairbanks of $18-an-ounce gold, of river steamers in summer and horse-drawn sleighs and dog teams in winter, and of tolerance of interdependence. We needed one another. The big world was far away.

But it was coming closer. In the middle of the morning on March 12, 1914, all the town's whistles suddenly started blowing. Immediately an impromptu parade formed down Front Street. The Alaska railroad bill had passed; James Wickersham, Alaska's one delegate to Congress, had triumphed. Highways were far into the future, but a railroad in the meantime would open both the Healy River coal deposits and Alaska to the world. In 1917, the construction campsite of Anchorage became headquarters for the ambitious enterprise to link Seward to Fairbanks, the northern terminus, 470 miles from the sea. A new era for Alaska had begun.

History is always a tale of inexorable changes, for man is such a restless creature. Fairbanks went from the prerailroad, prehighway, $18-gold period to a depression during World War I. Then came big mining companies with dredges and heavy machinery. World War II brought the military establishment. Later travel and tourism increased, for the airplane had made the entire state knowable.

Alaska in many ways, I think, was ready for the changes, eager for new ventures and activities. Yet Alaskans also cherished the independent spirit of the early years, and much of it endures to this day. On the Arctic coast the Eskimos have adopted many of the habits imposed on them by the white race, but they still subsist on the land, fishing and hunting whales, polar bears, caribou, seals, and birds. Prudhoe Bay and the Trans-Alaska Pipeline are only 50 miles west of the boundary of the Arctic National Wildlife Refuge, home of a 180,000-head caribou herd as well as many other mammal and arctic bird species. Millions of migratory waterfowl are protected in the deltas of the Yukon and Kuskokwim with the concerned cooperation of that region's natives. The largest of our national forests, the Tongass, is now being studied to determine how much logging is too much. Southwest of the "big city" Anchorage is the Kenai Peninsula with its moose refuge and offshore oil rigs which have been pumping away for about 30 years.

But overriding all of these manmade triumphs and concerns, we must always remember the "Great Country" and the still-victorious *power* of the land itself.

The Kobuk sand dunes, the untouched Noatak River Valley, Cape Krusenstern, Glacier Bay, Lake Clark, Togiak Refuge, and many more are now protected under the Alaska Lands Act of 1980. This act, one of the most important events in Alaska's history, established 10 new national monuments, preserves, and parks (one of which, the Wrangell-Saint Elias, is the largest national park in the United States), 9 new national wildlife refuges, 25 wild and scenic rivers — altogether 105 million acres.

It is a marvelous providential fact that, of all the areas chosen for national parks, wildlife refuges, and wild rivers, none holds minerals, oil, or timber to any

"I grew up in the Fairbanks of $18-an-ounce gold, of river steamers in summer and horse-drawn sleighs and dog teams in winter, and of tolerance of interdependence. We needed one another."

tempting degree. Thus we can be fairly sure that those 105 million acres will be there unspoiled into the future if they are given the kind of loving care which they deserve. Forty-nine conservation organizations are active in Alaska today; their members are people from every occupation. They are proud to live in Alaska. As my friend Joe Meeker said in a 1975 talk he gave in Anchorage, "Very few people come to Alaska by accident. It isn't easy to get here, and it isn't easy to stay. Strong values and beliefs are necessary simply to justify one's presence in this part of the world."

Granted that industry is here to stay, when all the nonrenewable natural resources have been dug up, piped away, or cut down, what will be left for Alaska? The one industry, aside from fishing, which can be most lucrative, nondestructive, and self-perpetuating for all time — a commodity in short supply in other world markets — the industry of simply letting people come, look, and enjoy!

I have talked to many tourists in Alaska and discovered that they are searching for a variety of things: vastness, magnificence, mountains, glaciers, great trees, whales, seals, birds, and other wildlife. They are also searching for glimpses of old Alaska, an informality of life, happy and enthusiastic people. I watched some sight-seers in Fairbanks stopping to look at a garden, admire the cabbages, the peas, and all the rest, and talk with the white-haired old-timer working in it. These are things travelers will remember, for people are always fascinated by people.

Today visitors to Alaska take home mental images of rich and free and innocent wild creatures. But will people 50 years from now be able to find, observe, and photograph birds and animals in their natural world as the Kaufmans have done for you in this book? Thanks to that Alaska Lands Act of 1980, I think they will, but I wonder about the forces now at work in this huge, indescribable "Last Treasure." Is there still that tolerance, that caring for one another that I grew up with in a time when so many treasures and pleasures were free, when great space and wilderness were taken for granted? They can be taken for granted no longer.

There may be people who feel no need for nature. They are fortunate, perhaps. But for those of us who feel otherwise, who feel something is missing unless we can hike across land disturbed only by our footsteps or see creatures roaming freely as they have always done, surely there should still be a wilderness. Species other than man have rights, too. Having furnished all the requisites of our proud, materialistic civilization, our neon-lit society, does nature, which is the basis for our existence, have the right to live on? Do we have enough reverence for life to concede to the wilderness this right? I submit that if our answer is "yes," then, when all the nonrenewable resources are gone, Alaska will still have one which can support a healthy economy and a happy life for her people for all time.

This is my hope and my prayer.

MARGARET E. MURIE

"There may be people who feel no need for nature. They are fortunate, perhaps. But for those of us who feel otherwise, who feel something is missing unless we can hike across land disturbed only by our footsteps or see creatures roaming freely as they have always done, surely there should still be a wilderness."

FIRST SNOWFALL, DENALI NATIONAL PARK AND PRESERVE

MOONRISE OVER MOUNT BROOKS

At the turn of the 20th century, gold drew thousands to Alaska. Although it was not the territory's first valuable natural resource, nor was it to be the last, its appeal was undeniable. Still some noticed other attractions. As Robert Service wrote:

> There's gold, and it's haunting and haunting;
>   It's luring me on as of old;
> Yet it isn't the gold that I'm wanting
>   So much as just finding the gold.
> It's the great, big, broad land 'way up yonder,
>   It's the forests where silence has lease;
> It's the beauty that thrills me with wonder,
>   It's the stillness that fills me with peace.

Increasingly in recent years, the Alaskan wilderness has been recognized as a resource in its own right, perhaps more precious than all the nuggets ever panned in the state's icy creeks.

Since 1980, when President Jimmy Carter signed the Alaska National Interest Lands Conservation Act into law, some 105 million acres have been set aside in new or expanded national parks, monuments, forests, preserves, and wildlife refuges. Of those acres, over half are designated as Wilderness where, according to the Wilderness Act of 1964, "the earth and its community of life are untrammeled by man, where man himself is a visitor who does not remain."

The quantity of these holdings is both staggering and misleading. To survive Alaska's long winters, a single moose may forage over 25 to 100 acres of land; a single brown bear on the Arctic Slope may range over 100 to 300 square miles. Alaska may be one-fifth the size of the lower 48 United States, but its short summers so limit vegetative growth that the land supports only a fraction of the wildlife that could live on the same acreage further south.

Located in the Alaska Range, Denali National Park and Preserve covers 6 million acres and is capable of sustaining entire ecosystems. Mount McKinley dominates the landscape. Early Athapaskan tribes called it *Denali*, "the great one," and its 20,320 feet make Mount McKinley North America's highest mountain, tall enough to create its own weather. Though in summer clouds shroud the twin peaks almost two days in three, the Athapaskans believed it was the home of the sun. No mountain on earth rises above its surroundings so precipitously, and no landmark in the state so strongly embodies the grandeur of what the Aleuts called *Alashka* — "the great land."

(ABOVE AND LEFT) DALL SHEEP, DENALI PARK   25

DALL SHEEP, DENALI PARK

TUNDRA POND; (RIGHT) BEAVER, DENALI PARK

ALASKA RANGE, DENALI HIGHWAY

32   (ABOVE AND RIGHT) CARIBOU, DENALI PARK

WRANGELL MOUNTAINS, WRANGELL-SAINT ELIAS NATIONAL PARK AND PRESERVE

When Secretary of State William H. Seward purchased Czar Alexander II's failing colony in 1867, a skeptical press dubbed Alaska "Seward's Ice Box." Few would claim that at two cents an acre the czar got the best of the bargain, but some misconceptions persist. Despite its image as a land of ice and snow, much of Alaska is green at least half of the year; glaciers cover only about three percent of the state. Because high precipitation is as essential to glaciers as low temperatures, most are located not in the far north, but rather in the rainy south-central and southeast regions.

Exploring Glacier Bay in 1879, John Muir wrote, "Here, too, one learns that the world, though made, is yet being made; that this is still the morning of creation." Glaciers build on mountain slopes, where countless heavy snowfalls become compacted in masses so dense that prismatic ice crystals reflect light as a deep "glacier blue." Drawn by gravity across the land like rivers towards the sea, glaciers carve U-shaped valleys in their grinding advance. Where tidewater glaciers retreat, inlets form.

Most Alaskan glaciers never reach the sea. Among the debris at the faces of those that do, like the 16 tidewater glaciers of Glacier Bay, are microscopic plants and animals attracting fish and sea birds. Harbor seals, sea otters, porpoises, and whales feed in the cold waters. Some of the makings of this glacial world are dramatic, as glaciers calve icebergs into the water so loudly that the Indians called the area "Thunder Bay." Others are much more subtle.

After a glacier recedes, mosses and lichen work on the rubble that remains, turning it into new soil. Over the course of 20 to 30 years, first small tundra plants, then alder, willow, and cottonwood take over. After some 100 years, western hemlock and Sitka spruce reclaim the land until another glacier advances over them. The lofty trees of the coastal rain forests are but part of an endlessly rhythmical process that begins with the falling of snow.

ICE FIELD, KENAI FJORDS NATIONAL PARK   37

MENDENHALL GLACIER, JUNEAU

JOHNS HOPKINS GLACIER, "CALVING," GLACIER BAY; HUBBARD GLACIER (RIGHT)

GLACIER BAY, GLACIAL TERRAIN; GLACIAL "BLUE ICE," ICE CAVE ENTRANCE

HIKERS ALONG TUSTEMENA GLACIER, KENAI NATIONAL WILDLIFE REFUGE

TOTEM POLE, GOVERNORS'S MANSION, JUNEAU

POD OF HUMPBACK WHALES FORM "BUBBLENET" TO CATCH HERRING

The geological forces that created the magnificent Alaska Range are still at work in Alaska. As the tectonic plates that make up the earth's thin crust grind past one another, the earth quakes. When magma, the molten matter upon which these plates drift, forces its way to the surface, volcanoes erupt. Because Alaska's southern coastline lies along the boundaries of two tectonic plates, seismic activity is inevitable. On the Alaska Peninsula are more than one-tenth of the world's known volcanoes; some of these are unquestionably active, though seldom so violently as Katmai, which in 1912 buried 40 square miles of valley under 600 feet of ash.

More than any other Alaskan mammal, the brown bear typifies the latent power of a land which, for all its beauty, can tremble and explode in a shower of volcanic ash. Because it can be fatal to startle a large animal capable of running up to 35 miles per hour, some hikers try to keep the wind at their backs; others attach "bear bells" to their packs or sing as they walk. Campers are warned not to pitch their tents near berry patches, bear trails, and food caches. If they inadvertently wander within the 50-yard radius which makes up the bear's "critical space," they are advised not to interpret its attempts to get a better look at them as an offer to charge, nor to encourage it to chase them by running.

Once thought to be separate species, the Interior's grizzly and the coastal brown bear are both now considered *Ursus arctos* distinguished primarily by their location and diet; the grizzly consumes fewer fish and more roots and berries. Despite the *horribilis* that was part of the grizzly's name, brown bears prefer fishing and hunting squirrels to killing larger animals, an often hazardous, futile undertaking. Near humans, they flee more often than fight, but they are unpredictable, warranting cautious respect. Of all Alaskan animals, the brown bear most reminds us that it is we who intrude in the wilderness.

DALL SHEEP, KENAI PENINSULA   51

COW MOOSE, KENAI PENINSULA

54   MARCH 1986 ERUPTION; (RIGHT) DOME FORMING BEFORE AUGUST 1986 ERUPTION, AUGUSTINE VOLCANO

BROWN BEARS, MCNEIL RIVER

60   (ABOVE AND RIGHT) BROWN BEARS WITH SALMON, ALASKA PENINSULA

LOW TIDE, MCNEIL RIVER

BALD EAGLES, KENAI PENINSULA

WALRUS, TOGIAK NATIONAL WILDLIFE REFUGE

Each year thousands of walrus clamber upon ice floes with the help of their heavy ivory tusks, feed on clams dredged from the bottom of the sea, and swim along Alaska's coast to summer in the southern Bering Sea. For many other species, in an environment of limited resources and dramatic seasonal changes, migration is also essential to survival.

Over 20,000 years ago the Asiatic ancestors of Alaska's Indians followed migrating herds across the land bridge over the Bering Sea into Alaska. Traditionally the welfare of native Alaskans has been linked to the waxing and waning of animal populations; elderly Yup'ik Eskimos in the Yukon Delta, the water-logged flatland between the Yukon and Kuskokwim Rivers, recall times when the hunting was bad and starvation drove them to eat boiled mice and the leather of their mukluks. Birds and mammals still play an important role in the lives of Alaskans. Under the terms of the Alaska Native Claims Settlement Act of 1971, the native inhabitants of small villages throughout the Yukon Delta National Wildlife Refuge, like those in other Alaskan wildlife refuges, continue to trap, fish, hunt, and gather eggs, maintaining their traditional way of life.

During the summer, some 170 migratory bird species flock to the rivers and ponds of the Yukon Delta. All of the world's cackling Canada geese, smaller than their better-known cousins, with a shriller, brittle call, and 90 percent of its bristle-thighed curlews nest here among clouds of mosquitoes. Black scoters and common loons, emperor geese and black brant, spectacled eider and long-tailed jaegers, all share the nation's largest river delta with moose, wolves, beaver, and muskrats. The birds come from as far away as the southern coast of Chile, adding their cries to the honking of geese that signals the end of the long, dark winter to the Eskimos of the Yukon Delta.

COMMON MURRES, TOGIAK BAY

(LEFT) WALRUS, CAPE NEWENHAM; WALRUS, CAPE PEIRCE   71

74   CACKLING CANADA GEESE, YUKON DELTA

AERIAL VIEW; (RIGHT) BLACK BRANT, YUKON DELTA

(LEFT) RED-THROATED LOON; COMMON LOON, YUKON DELTA 79

80   SPECTACLED EIDER, YUKON DELTA

(ABOVE AND RIGHT) TUNDRA SWANS, YUKON DELTA

SUNSET AT MIDNIGHT ON THE SUMMER SOLSTICE, TUTAKOKE RIVER

Lured by the Bering Sea's protein-rich waters, millions of migratory sea birds jostle for perches and lay their eggs on rocky ledges in the cliffs of the Pribilof Islands. Below are the rookeries where the world's largest herd of northern fur seals gathers to mate and bear young, each returning yearly to the place of its own birth. Squawking birds and bawling seals seem as intrinsic as the winds that blow across these fog-bound islands 300 miles from the Alaska mainland. But in an environment where loud noises can panic an entire bird colony, causing them to knock hundreds of eggs and new chicks into the sea, existence is a question of balance.

When survivors of the expedition on which Dutch explorer Vitus Bering died returned to Siberia in 1741, the sea otter pelts they brought with them roused such a desire for furs that hundreds of Russian hunters and traders thronged to Alaska. By the early 1900s, both the sea otter and fur seal were nearly extinct. But sea mammals were not the only victims of this exploitation. The ancestors of the Aleut-Russians who now live in the Pribilof Islands were brought there to slaughter fur seals for the Russians. It is estimated that some 10,000 Aleuts died during this period of virtual enslavement.

The musk-ox herds of Nunivak Island attest to an earlier collision between human desires and the natural world. When threatened, musk oxen gather in a circle or huddle near cliffs, turning their horns towards the danger. This instinct, which helped them fend off wolves for centuries, proved useless against men with guns, and the last of Alaska's original musk-ox population were killed by Eskimo hunters before the turn of the 19th century. Like the reindeer brought to Alaska in the early 1900s, Nunivak's musk oxen are the descendants of animals imported from Greenland. Today, prized for their thick undercoats of *qiviut*, a cashmere-like wool, these rare animals wander the island's windswept coastal dunes, living remnants of the Ice Age when they ranged throughout much of North America.

TUFTED PUFFIN, PRIBILOF ISLANDS   87

(ABOVE AND RIGHT) ARCTIC FOXES, PRIBILOF ISLANDS

STELLER'S SEA LIONS, NABANGOYAK ROCK

REINDEER, NUNIVAK ISLAND

COTTON GRASS, NUNIVAK ISLAND

102   SALMON FISHERMEN, NASH HARBOR

So little precipitation falls above the Arctic Circle that the area would be a desert were it not for water locked below ground in permafrost. Underlying more than two-thirds of the state, this layer of frozen subsoil, gravel, and rock is as much as 2,000 feet thick in parts of the North Slope between the Brooks Range and the Arctic Ocean. Each spring the permafrost's top layer thaws just enough to sustain plants and the animals that feed on them.

In June the arctic tundra blooms with the tiny flowers of plants that have developed ways to deal not only with cold winters but also with a growing season of singular intensity, rarely longer than nine weeks or warmer than 45°F. Their stunted size helps protect arctic plants from incessant winds. In some varieties waxy leaves slow the evaporation of moisture, and in others hairy stems or clustered buds serve as insulation. Many arctic plants reproduce asexually; those that don't often have unusually hardy seeds, thickly encased. Arctic lupine seeds have been known to sprout after lying dormant for nearly 10,000 years.

Despite its rugged appearance, the arctic tundra is extremely vulnerable. Because permafrost prevents efficient drainage for its thawed top layer, the water that pools in a misplaced footprint may remain there for years, scarring the land. This harsh environment makes uncommon demands on its wildlife. Only the ceaseless migrations of the caribou herds, summering on the North Slope and wintering south of the Brooks Range, enable thousands to graze on sparse, slowly growing vegetation without depleting it. With their acute sense of smell and wide front hooves, caribou can locate and dig for lichens buried under the snow.

Arctic birds have also adapted to this environment. White in winter, mottled in summer, the ptarmigan's changing plumage helps protect it from predators. In a treeless landscape, the snowy owl nests on mounds of matted grass to keep watch for foxes. One of the few birds to live on the North Slope year-round, it hunts by day as well as by night, an essential adaptation to summers when the sun does not drop below the horizon for as many as 84 days.

106   BROOKS RANGE; (RIGHT) WILLOW PTARMIGAN, NORTH SLOPE

MIGRATING CARIBOU, KONGAKUT RIVER

(LEFT) CARIBOU ANTLER AND DOUGLASIA ARCTICA; CARIBOU PASS, NORTH SLOPE   113

114   (ABOVE AND RIGHT) LUPINES, NORTH SLOPE

SNOWY OWL, ARCTIC NATIONAL WILDLIFE REFUGE

BULL MOOSE, AUTUMN TUNDRA, DENALI NATIONAL PARK

SHORT TAILED WEASEL, ARCTIC GROUND SQUIRREL, HOARY MARMOT. DENALI NATIONAL PARK   119

BLACK BEAR TRACKS, NUKA ISLAND, KACHEMAK BAY STATE PARK

122   BLACK BEAR, KENAI PENINSULA

124   SAND PATTERNS, OUTGOING TIDE, NUKA ISLAND, KACHEMAK BAY STATE PARK

126   ORCA LEAPING, CHASING DALL PORPOISE PREY JUST AHEAD

NUKA ISLAND, COASTAL OLD GROWTH, SITKA SPRUCE FOREST

YELLOW LADYSLIPPER, SOUTHEAST ALASKA; WHITE BOG ORCHID, WOOD FROG, DENALI NATIONAL PARK; TINY LICHENS, ALPINE TUNDRA, LAKE DOROTHY, TONGASS NATIONAL FOREST

ALASKA RANGE, ALPENGLOW

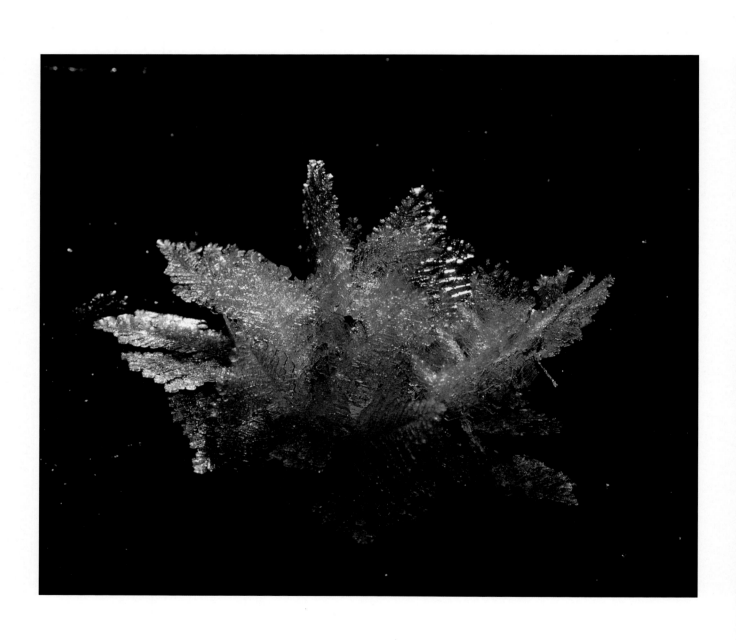

132   FROST FLOWER, ICE CRYSTALS, RAVINE CREEK. MINUS 50°F.

Diversity! Add to that, adversity, and one spells Alaska. Adjectives and categories range a full gamut of contrast, depending upon what is being viewed, from tiny, even microscopic to huge, towering or in the lexicon of a young generation, perhaps "humongous!" The tallest of mountains gives way to flatness of tundra or tidal deltas, the smallest of insects might cause more concern than the largest of bears, and gaudiness of puffins is a foil for camouflage of ptarmigans or most wildlife. While sitting at breakfast on uninhabited Nuka Island, one might watch the antics of a tiny shrew scampering under spruce roots, while in the cove otters romp and humpback whales breach, providing the long and short of the animal kingdom.

Alaska *is* the long and short, the heavy or light, volcanic heat and arctic cold, daylight and long darkness, the wet or dry. Its schools of fish, flocks of birds, or herds of caribou which liken it to a Serengeti of North America, oppose images of the solitary eagle, wolverine, grizzly, or moose. Rage and ferocity of a half-ton brown bear charging an intruder or prey softens to tender and caring behavior shown its cubs by a grizzly mother. Abundance of natural riches are offset by the adversity of winters and terrain for the human exploiter — who finds that the largesse can be rewarding, or, alternatively, might kill him.

Visitors who hike up any number of small swift streams, "flightsee" over peaks in the Kenai fjords, or who drive by acres of mine tailings gasp at the degree to which miners drove themselves and their tools — animal, human, or material — in wresting minerals, especially the elusive gold, from mountains of quartz or mud. Those resourceful and tough miners opened Alaska much as they did the western mountain states. Recent years have shifted to "black gold" as the primary mineral target, with continued lumber and fishing weighing in as impacts in "taming" the huge state. Not to be ignored, and possibly to become a principal causative in opening up previously inaccessible areas, are great and growing numbers of tourists who revel in sampling the rich history of Russian and native legacies or experiencing the diverse wilderness as they fly or boat to fantastic fishing or hunting spots or hike delightedly in the shadow and autumn splendor of Mount McKinley.

SANDHILL CRANE MIGRATION, HOMER AREA, AUTUMN

136   GRAY WOLF, ARCTIC NATIONAL WILDLIFE RESERVE

WILLOW LAKE, WRANGELL-ST. ELIAS NATIONAL PARK; MERLIN FLEDGLINGS (RIGHT), DENALI NATIONAL PARK

MUSKOX HERD, NUNIVAK ISLAND

WILLOW PTARMIGAN (TOP), WINTER PLUMAGE, RAVINE CREEK; WALRUS ON ICE FLOE, CAPE NEWENHAM   143

AUTUMN TUNDRA, SAVAGE RIVER AREA, DENALI NATIONAL PARK

ALASKA — A State of Mind! Truly, it is that — a place where the earth trembles frequently as tectonic plates shift beneath its magnificent surface, an untamed wilderness of forests, mountains, and streams — a place where numbing cold of a winter midnight provides a stage for the crackle of northern lights as the skies treat us to a shimmering kaleidoscopic veil of greens, whites, and lavenders. One of the world's least populated areas, it is a place where roads reach less than one-fourth of the state, where moose-car or moose-train collisions are a real hazard for road or rail, where tourists in Homer gawk at the sight of a moose appearing at the takeout window of McDonald's, and whose largest city features the busiest airport in the world — for float planes!

I return again to Alaska as a photographer, a role I've played annually for two decades. During those visits I have truly fallen in love with the land and its inhabitants, a love which one might feel is at odds with a career in our submarine Navy, where I was a leading advocate for the massive and costly defense systems which history shall acknowledge deterred a war of unimaginable horror. I offer this to rebut any idea that I am what developers derisively term a "tree-hugger," though I find that I associate myself increasingly with many of their concerns. In truth, prior to my underseas career I was an ardent, serious student of nature and its wonders, though I pursued hunting and fishing in a major way. In my dotage I have a growing awareness that it is the wonders of Alaska which nudge me back to the softer, more gentle boy who loved the woods, knew all the birds and plants, and sired a son who never left that early delightful stage. As I discuss environmental issues with people in all walks of life, I constantly seek to frame my concerns in words which will convey accurately the wonder, the beauty, the importance, and the need of this remarkable area — and my fears that forces of politics and greed, fueled by that product of our conditioning — progress — increasingly threaten Alaska.

Most serious Alaska photographers, as with practicing naturalists, can cite occasions where their wildlife subjects seemed to accept their nearness, almost embracing them as one of their groups or in their behavior. Compared to many colleagues my experience places me in a ranking of an apprentice — a pretender. Still, over years I have had a number of encounters which were wonder-provoking, amusing, strange, or simply puzzling. On one occasion a normally dangerous cow moose with calf tolerated my slow, patient stalk and subsequent close presence for hours. I felt "accepted" when first the calf, then the cow, settled down for a nap as I snapped photos, even portraits, while hearing clearly the loud rumbling of their stomachs. At another time a sprightly red fox appeared at our tent on the Alaskan peninsula , seemingly showing off to us his mouthful of lemmings and birds. Several times he buried his prey only yards away, turning to us as though with pride and seeking approval before he dashed off for a refill.

148   RED FOX, LEMMINGS AND BIRDS IN MOUTH, HALLO BAY, KATMAI NATIONAL PARK

I suppose my favorite photo subjects are bears, but I retain a cherished spot for the curled-horn Dall Sheep rams which have given me great satisfaction. Not only are they beautiful subjects, but they also are found at heights favoring spectacular scenes as well as providing a physical workout in getting to them.

Recently I recounted my experiences to a close friend of mine who was about to visit Alaska for the first time. I spent long minutes detailing the arduous climb up a mountain slope of scree to find a ram which my son, Steve, told me had been sighted — "and Dad, he has one blue eye, or else a cataract!" An hour's climb was rewarded, as, after a slow step-by-step stalk I seated myself on the fringe of a group of eleven large rams, each peacefully chewing its cud as I worked film and camera and looked, to no avail, for a blue eye. For several hours I was as one of the herd, nibbled a chocolate bar lunch, avoided sudden moves, and listened to a chorus of growling stomachs as I composed photos. One by one a ram would rise, chew at grass, stretch a bit, "pose," and lie down again. Finally the largest ram, closest to me, arose, languidly stretched, and for the first time, turned to face me. A blue right eye! I was thrilled. My telephoto lens showed clearly a scar over the damaged eye — this herd leader's badge of combat. Some unheard signal was communicated, perhaps

by body language, but one by one the rams arose, stretched, and following some prehistoric protocol joined up single file in descending order of horn size, following "Blue-eye" as the band climbed to ledges on the highest cliffs — an aging photographer bringing up the rear.

I could not express fully the marvelous acceptance by the rams, the mystery of their communications and the obvious leadership displayed, and the closeness I had felt. I exclaimed "Just totally absorbing, fantastically interesting!" My friend was wistful as she offered that my words were inadequate, and I saw that her eyes were filled as she spoke, " I think the right word is 'spiritual!'"

Today I look upon a soft, misty Kachemak Bay scene which virtually whispers, and I join those who urge moderation in what is called progress — who hope to preserve a relatively untamed Alaska. I would add an adjective to a brimming lexicon which includes "magnificent," "inspirational," "spectacular," "breathtaking" — and thousands more: ALASKA — Spiritual!

R. Y. "Yogi" Kaufman

ALPENGLOW AT DAWN, MOUNT MCKINLEY

CAMPERS, LAKE DOROTHY, TONGASS NATIONAL FOREST

ALASKA PIPELINE, NORTH SLOPE, LATE AUTUMN; (RIGHT) FLY-IN FISHING, CRESCENT LAKE, LAKE CLARK NATIONAL PAR

ALASKA RANGE, DENALI PARK

CATHEDRAL SPIRES, DENALI PARK